Health 127

你准备好了吗？

Are You Prepared?

Gunter Pauli

冈特·鲍利 著

凯瑟琳娜·巴赫 绘

章里西 译

特别感谢以下热心人士对译稿润色工作的支持：

王必斗　王明远　王云斋　徐小怙　梅益凤　田荣义
乔　旭　张跃跃　王　征　厉　云　戴　虹　王　逊
李　璐　张兆旭　叶大伟　于　辉　李　雪　刘彦鑫
刘晋邑　乌　佳　潘　旭　白永喆　朱　廷　刘庭秀
朱　溪　魏辅文　唐亚飞　张海鹏　刘　在　张敬尧
邱俊松　程　超　孙鑫晶　朱　青　赵　锋　胡　玮
丁　蓓　张朝鑫　史　苗　陈来秀　冯　朴　何　明
郭昌奉　王　强　杨永玉　余　刚　姚志彬　兰　兵
廖　莹　张先斌

目录

Contents

一只蜜蜂和一只鸽子正在讨论他们未来可能会遇到的“飞来横祸”，商量该如何预防这些灾害。

“我能飞越任何区域，把信息或地图捎给朋友来帮助他们。”鸽子骄傲地说。

A bee and a pigeon are talking about any nasty surprises they may have to deal with in the future, and how they can prevent bad things from happening to them.

"I can fly across any zone and take messages or maps to help our friends," Pigeon proudly shares.

一只蜜蜂和一只鸽子正在讨论……

A bee and a pigeon are talking ...

我能用鼻子闻出炸弹……

I can sniff out bombs ...

“没错，你送情报送得特别准。” 蜜蜂羡慕地说。

“那你有什么特长能帮到我们的朋友吗？”

“我能用鼻子闻出炸弹。”

“Yes, you are very good at getting information into the right hands,” Bee says, admiringly.

“And what can you contribute to help our friends?”

“I can sniff out bombs.”

“真的吗？我只知道狗可以闻到危险的气息。如果你也可以，甚至能比狗做得更好，那真是个大惊喜！”

“我们蜜蜂经常给人带来惊喜。他们花了好多年才发现蜂蜜、花粉、蜂蜡和蜂王浆的好处。他们现在发现，我们的嗅觉也很灵敏。”

“灵敏到可以‘扫雷’？”

“Really? I know dogs can sniff out danger, but that you can do so too, perhaps even better than them, comes as a big surprise.”

“We bees always surprise people. It took years for them to discover the good things in our honey, pollen, wax and royal jelly. And now they have figured out that we are great sniffers.”

“But sniffing out bombs?”

我们的嗅觉很灵敏……

We are great sniffers ...

……感知散播在空气中的花粉……

... sense pollen in the air ...

“其实，我们的嗅觉与生俱来。我们的触角可以感知散播在空气中的花粉，然后追踪到对应的花朵。我们还能嗅出糖，在找到糖之后，我们会通过跳‘摇摆舞’把信息传递给朋友和家人。”

“闻出花粉和找到炸弹可是两回事，我的朋友。”

“Well, we are natural-born sniffers. With our antennae we can sense pollen in the air, and track it down to specific flowers. We can also sniff out sugar. When we find it, we do a waggle dance to relay the information to our friends and family.”

“There is a big difference between smelling pollen and finding bombs, my friend.”

“如果把危险物品的气味与糖混合，我就可以通过训练辨别出危险品的气味，然后跳摇摆舞。”

“你的意思是人们只要看到你的摇摆舞，就会明白有危险了？”

“呃，如果只有一只蜜蜂跳舞，没有人会在意。但是当我们一起舞动和摇摆，人类就会意识到可能有地雷埋在地下了。”

“By mixing sugars with the scent of things that are dangerous for people, I can be trained to pick up the odour of explosives. Then I do my waggle.”

“You mean that seeing you waggle is all people need to figure out that there is danger?”

“Well, if one of us does it no one is impressed. But when we all start winging and waggling, people know that there may be a landmine buried in the ground.”

……然后跳摇摆舞……

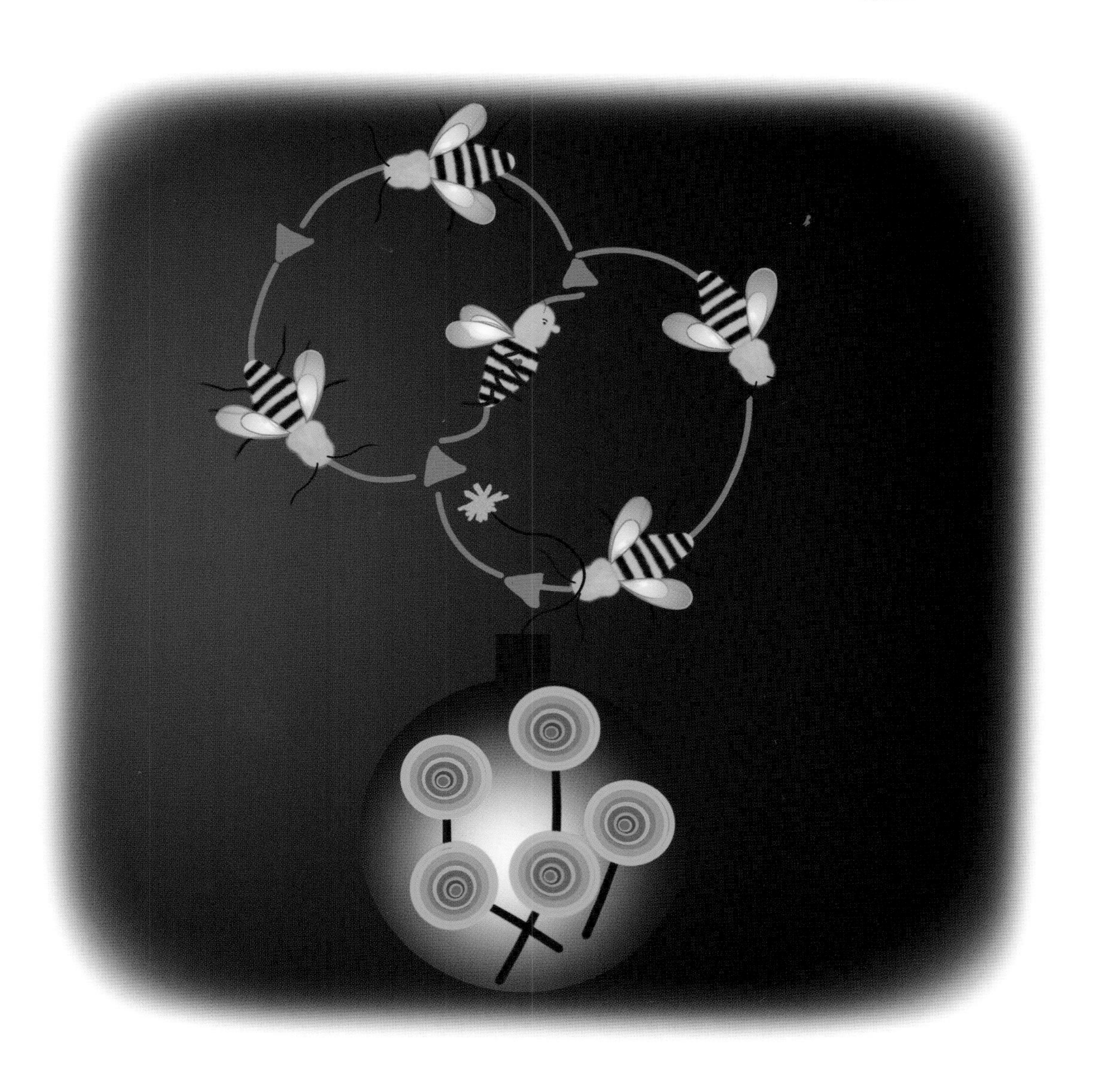

… then I do my waggle …

我们还有军人朋友……

We have our friends in the military ...

“人们能在爆炸前就把炸弹找到，这真是太好了。”鸽子说。

“是啊，我们还有军人朋友做后盾，竭力保护我们免受伤害。”

“我只希望人们在日常生活中对待他们的健康以及环境也这么认真就好了。”

“It is good to know that people can find these before they explode,” Pigeon says.

“Yes, fortunately we have our friends in the military who will do everything in their power to protect us against harm.”

“I just wish that people would do the same in their daily life when it comes to their health and our environment.”

“哎，说得没错！”蜜蜂感叹道，“太多人的生活方式不健康，得病后才求助医生。而且很多人根本意识不到他们的行为对大自然造成的不利影响。”

“得病当然要去看，不过我们能为预防疾病做点什么吗？”

“Oh absolutely!” Bee agrees. “So many people do not live healthy lives, and rely on doctors to provide a cure when they get ill. And so many people are not even aware of the negative impact their actions have on Nature.”

“Of course you need medicine when you are ill, but what can you do to prevent illness?”

我们能为预防疾病做点什么吗？

What can you do to prevent illness?

慢慢地破坏环境……

Steadily destroying the environment ...

“问得好！为什么一定要等到来不及了才开始治疗？防病难道不比治病更重要？”

“预防疾病十分有意义。”鸽子答道，“但你得清楚地知道你要预防的是什么。”

“同样地，如果我们意识到环境在慢慢被破坏，我们要等到它坏得无法修复了才去改变我们的行为吗？”

“Good question! Why wait until it is too late to take a remedy? Is prevention not better than cure?”

“It does make sense to rather prevent illness,” Pigeon replies. “But you have to know what it is you want to prevent from happening.”

“And when we know that we are slowly and steadily destroying the environment, do we wait until Nature has been damaged beyond repair before we correct our behaviour?”

“当然不，现在我们就得行动起来！”蜜蜂说。

“没错！是时候像军队一样思考了！我们得确保已经做好充分准备。”

……这仅仅是开始！……

“Of course not, we must do so now!” Bee says.

“Exactly! It is time we all start thinking like the military, and ensure that we are prepared well in advance.”

... AND IT HAS ONLY JUST BEGUN!...

……这仅仅是开始！……

...AND IT HAS ONLY JUST BEGUN!...

Pigeons were the first birds to be domesticated. Pigeons know where their home is and when released from a distance of up to 1000 kilometres away, will return home, flying at a speed of 60 km/hr.

鸽子是人类最早驯化的鸟类。它们能记得巢穴位置，即使放飞到 1000 公里以外，也会以每小时 60 公里的速度归巢。

The Romans used carrier pigeons during military campaigns and Julius Caesar deployed pigeons in his conquest of the Gaul. The Greeks used carrier pigeons to inform people in other cities of the winners of the Olympic Games.

罗马人在军事行动中已使用信鸽，恺撒在征服高卢期间就曾部署信鸽传递信息。希腊人用信鸽告知其他城市市民关于奥运会优胜者的信息。

Croatian scientists unveiled research into using colonies of bees that can detect land mines from more than 5 kilometres away. The bees are trained by being fed a sugar solution mixed with the smell of explosives.

克罗地亚科学家公布了一项利用蜂群探测 5 公里外地雷的研究。科学家用混有炸药气味的糖溶液饲养蜜蜂以训练其探测能力。

Bees can smell so well that they know all about a substance from the scent alone. This enables them to 'taste' a substance before it even touches their tongues.

蜜蜂的嗅觉极其灵敏，单凭气味就可以辨别出物质。这让它们不用口器触碰就能“尝到”物质的味道。

Dogs show signs of distress before seismic activity, which they perceive either through their exceptional hearing (of rocks crumbling under the Earth's surface), or through sensing seismic activity through their paws.

狗在地震前会显得焦躁不安。它们能利用超凡的听觉听到地表下岩石破裂产生的次声波，或通过爪子感知地震性活动。

Through their sense of smell dogs can predict the onset of labour in pregnant women. They can also detect the faint odours of an illness in people suffering from diseases such as cancers and diabetes.

狗能够通过嗅觉察觉孕妇是否即将分娩。它们还能嗅出癌症和糖尿病等患者身上微弱的气味。

If the military were to wait for scientific proof that an enemy will attack, it will be too late to adequately prepare for defence. The military must study all risks, assess all options and take precautions – well before proof is delivered.

如果军队真要等到有证据表明敌人会发动攻击后才行动，则会来不及做好充分的防御准备。军队必须在掌握充足证据之前分析所有风险，列出所有方案，从而防患于未然。

Some policy makers want proof that climate change is indeed occurring and want to know the magnitude of it, before formulating policies to mitigate human impact on the environment. But by the time proof is delivered, it will be too late to protect people against it.

在制定政策以减小人类活动对环境的影响之前，一些决策者想证明气候变化的确正在发生，并想进一步了解变化程度。但是当证据确凿时，对抗气候变化为时已晚。

Think About It 想一想

Who would you trust more to give you information about your health or safety: a pigeon, a bee or a dog?

如果有人给你一些关于你的健康或安全的情报，你会更信任谁：鸽子？蜜蜂？狗？

Would you wait until the enemy attacks, before preparing for the defense of your country?

你会等到敌人进攻时，才开始为祖国筑起防线吗？

What is the best option to stay healthy: to prevent illness or to have the right cure available when you do get ill?

怎样才能更好地保持健康？是预防疾病，还是等到生病后再治疗？

Are you forewarned when one person tells you about a danger, or do you need several people to tell you that trouble is ahead, for you to believe it?

一旦被告知有危险，你就会提高警惕，还是要等到许多人告诉你时你才会相信？

How would you get a message to your friends on the other side of the city, if you had no phone or internet? What other ways and means are there to invite friends to come and play or to go and see a film, or to warn others of danger? What are some of the quick and easy ways to establish contact without modern communication techniques? When you find one, test it and make sure it works. Ensure that it is fast and reliable.

假设没有电话或者互联网，你会怎么把消息传递给在城市另一头的朋友？假如你想邀请朋友来家里玩，或者和朋友外出看电影，或者警告他人有危险，除了现代化的通讯科技，有没有其他简便快捷的办法与他人取得联系？如果你发现了某种方法，亲自尝试一下，确保这种方式有效、快捷、可靠。

学科知识

Academic Knowledge

生物学	信鸽能通过耳内的传感器感知地球磁场，从遥远的地方找到回家的路，还可以根据太阳的位置判断方位；鳟鱼能利用鼻子内的传感器寻找适合繁殖的地点；赤蠵龟能够辨别自己所在地的经度和纬度；蜜蜂的触角能获取嗅觉信息，其高度灵敏的绒毛能够感知震动；鸽子是“一夫一妻”生活的鸟类，它们非常关心后代及种群其他成员；狗经过训练可以嗅出地雷内爆炸物的气味。
化学	蜜蜂触角内的170多个化学感受器让它在飞行中也能感知气味；蜜蜂能通过嗅觉获取更丰富的信息，不需用口器就可以判断某种物质的成分；化学品可以用于排雷：一种化学制剂可以让引爆装置失灵，另一种制剂可以将地雷和附近土壤固化，使专业人员可以将其安全地移除。
物理	鸽子能像GPS定位一样辨别地球磁场的极性和强度；蜜蜂通过细密的绒毛感知周遭的震动频率，异常的震动可能触发预警。
工程学	扫雷：数以万计的地雷是先通过蜜蜂和狗进行初步检测后再人工排除的；地雷埋入地下50年后仍能保持破坏力；由于地雷中塑料比例越来越高，金属成分越来越少，金属探测器已经很难探出地雷；探地雷达可以有效地定位地雷；军队传统上会运用所有可用资源进行模拟演习，评估针对不同人群和环境风险应制定何种防御机制。
经济学	保险业需要人们有采取预防措施以降低各种损害发生的风险意识；在气候变化的情形下不作为的成本：政府只修复已有的破坏，还是进行投资、预防新灾害的发生（如海平面上升）。
伦理学	在我们得知风险很高，而得到证据再行动将为时过晚的情况下，不能等获得了确凿证据再采取预防措施；战争中地雷的使用（通常是引起大多数平民伤亡的原因）使人们在战后50余年仍饱受安全问题困扰。
历史	鸽子在欧洲和中东地区作为主要肉食用鸟已有几千年的历史，而在印度和亚洲，人们可能更倾向于食用家禽；1972年，罗马俱乐部发表了研究报告《增长的极限》，利用数学对未来世界进行预测。
地理	20世纪90年代，克罗地亚因战争被埋入一百多万颗地雷；当地科学家设计了蜜蜂摇摆舞检测系统，用于探索未爆地雷，减少人员伤亡。
数学	对预防性维护的投资和收益进行量化后，我们可以确定预防比修复更为便宜；非线性的计算机模型（如系统动力学）通过反馈回路及乘数效应，能够把可能出现的变化对其他变量产生的影响考虑在内。
生活方式	社会常常会忽视一些提示灾害的先兆，不对风险进行评估，只有出现可信的证据时才要求政策制定者付诸行动。
社会学	鸽子是和平的标志；在中国，鸽子象征着信义与和睦。
心理学	除非迫于无奈，人类一般会避免做出艰难的决定——为什么大多数人只有在得知患癌后才开始戒烟?
系统论	做出采取预防措施或进行投资等决定以保护社会或家庭之前，我们要进行仔细评估和反思。在这些评估中我们需要研究多种因素间的交互关系、分配到各个因素的风险以及一系列关于机制的假设。

情感智慧

Emotional Intelligence

蜜　蜂

蜜蜂从对话一开始就表现出了对鸽子的敬佩，然后用平和的语气跟对方说自己也能做很了不起的事（嗅出炸弹）。蜜蜂首先陈述了众所周知的事实，随后补充了一些鲜为人知的新知识（嗅探）。他准备好了如何解释背后的机制，并用简单的术语详细解释了蜜蜂嗅觉灵敏的原因。他传达信息的方式十分清楚，动员蜂群做同样的摇摆舞动作。蜜蜂具有分析性思维，通过赞同鸽子的观点显示出共情。之后他运用了更为哲学性的思维方式，提出预防比治疗更有效的观点。

鸽　子

鸽子为自己的成就和能力感到自信和骄傲。他仔细思考了蜜蜂承担的角色及重要性，通过询问蜜蜂“你能做什么”，可以看出鸽子认为自己比蜜蜂更重要。得知蜜蜂的嗅探能力后，鸽子非常惊奇，并反复询问蜜蜂以再次确认。当鸽子相信了蜜蜂的独特性后，他开始了哲学思考，表示在有证据表明攻击即将来临或自然灾害可能发生前，做好充分防御是十分必要的。鸽子从保护领土的思考进一步拓展到对人类健康的关注，对人们不能及时采取预防措施表示担忧。

艺术

The Arts

一起来欣赏来一些最著名的鸽子画作，比如这些出自两位20世纪的艺术家之手的作品。首先欣赏毕加索在《和平鸽》中运用的简单线条，随后欣赏勒内·马格里特在《春天》中创作的蓝天白云鸽子图。你认为这些画家为何会使用鸽子的形象？艺术家们想传达给我们什么信息？

思维拓展

Systems: Making the Connections

世间充满了美妙的惊喜，同时也充斥着许多风险。我们拥有的资源很有限，而当某些风险转变为真实的灾害后，展开行动通常为时已晚。因此我们经常会意识到，无论采取何种预防措施，都不足以应付问题的严重性。日本多年来通过建造墙体和抗震建筑以抵御地震和海啸。然而当2011年3月11日一场巨大的海啸侵袭日本海岸之时，这些防御措施并没能阻止巨浪吞没海岸线，近两万人遇难。军队可能是社会中唯一有组织地时刻准备抵御外来风险的团体。国防资源是有限的，而外来攻击或自然灾害的来袭往往突如其来。军队需要对可用资源的分配进行权衡，判断该如何减少潜在的危害。计算机模拟技术应运而生。它可以分析一系列需要考量的关键因素，其中包括人口爆炸、自然资源的匮乏、气候变化、失业、污染、食品安全问题以及可用淡水资源短缺等。计算机模拟技术可用于研究单因素变化对其他因素造成的影响，将数千变量转化为方程式，通过反馈回路形式进行交互关联。这就导致了乘数效应，这些效应的好坏不一。在应用不同的假设的情况下通过改变我们的应答策略可以看到结果的动态变化，这些变化有时候可能与直觉背道而驰，然而却能更加真实地反映我们面临的挑战。尽管军方大规模使用了此种模拟技术，但其在民用领域的应用却十分有限，特别是对于气候变化所产生影响的预测则更为缺乏。使用动态模拟技术能够让社会能更加从容地决定哪里需要投资，什么需要改变。毕竟我们知道气候正在发生变化，但并不清楚每一个相关因素将具体如何变化。然而，如果社会要等到证据确凿后再付诸行动，就太迟了。

动手能力

Capacity to Implement

想象一下，50年内海平面将上升一米，100年内将上升两米，列出这些改变会对你的国家海岸地区造成的损害，然后对需要优先解决的问题列表。自己想想谁是预防和保护措施实施的监督者，而实施者又是谁；最重要的是，想想谁来支付这笔巨额投资所需的费用。然后就“如何尽快开始实施我们一致认为需要的措施”展开辩论。

故事灵感来自

杰伊·弗瑞斯特
Jay Forrester

当杰伊·弗瑞斯特用旧车零件组建了第一个12伏电气系统时，他还只是个住在父亲农场里的十几岁的孩子。之后他成了一名电气工程师，并掌握了计算机工程技术。他终生都在麻省理工学院斯隆管理学院工作。他提出了系统动力学模型，用于研究企业决策和城市发展动态等。此外，该模型表明，反直觉的政策往往会产生惊人的效果。1971年，弗瑞斯特起草了世界社会经济系统动力学模型的初稿。该模型反映了世界人口、工业生产、污染、资源及食物等因素之间的交互关系。该模型预测，如果不采取措施以减少人类对地球的需求，那么在21世纪的某个时候，世界社会经济系统将会崩溃。该研究促成了罗马俱乐部第一篇报告——《增长的极限》的诞生。

图书在版编目（CIP）数据

你准备好了吗？：汉英对照 /（比）冈特·鲍利著；（哥伦）凯瑟琳娜·巴赫绘；章里西译．—上海：学林出版社，2017.10

（冈特生态童书．第四辑）

ISBN 978-7-5486-1251-3

Ⅰ．①你… Ⅱ．①冈… ②凯… ③章… Ⅲ．①生态环境－环境保护－儿童读物－汉、英 Ⅳ．① X171.1-49

中国版本图书馆 CIP 数据核字（2017）第 143417 号

著作权合同登记号　图字 09-2017-532 号

冈特生态童书
你准备好了吗？

作　　者—— 冈特·鲍利
译　　者—— 章里西
策　　划—— 匡志强 张　蓉
特约编辑—— 隋淑光
责任编辑—— 汤丹磊
装帧设计—— 魏　来
出　　版—— 上海世纪出版股份有限公司学林出版社
地　址：上海钦州南路 81 号　电话／传真：021-64515005
网　址：www.xuelinpress.com
发　　行—— 上海世纪出版股份有限公司发行中心
（上海福建中路 193 号　网址：www.ewen.co）
印　　刷—— 上海丽佳制版印刷有限公司
开　　本—— 710×1020　1/16
印　　张—— 2
字　　数—— 5 万
版　　次—— 2017 年 10 月第 1 版
2017 年 10 月第 1 次印刷
书　　号—— ISBN 978-7-5486-1251-3／G·477
定　　价—— 10.00 元